PRACTICAL BOOK
FOR DRONES

How to Installed,

operation and

repairs drones

Matilda X. Kennedy

Table of Contents

CHAPTER ONE

INTRODUCTIONS

Drones are use for different purposes and serves as a supplementary security. Many drones have cameras, which allow you to see things from the drone's perspective. The digital can moreover file videos, so you can share your flying adventures with one of a kind people.

Every drone has a built-in flight controller that continues it stable. If a gust of wind pointers it over, the flight controller will proper away adjust the propeller speeds to re-level it. This makes it much less challenging for inexperienced persons to lookup to fly. Some drones have greater wise factors that let them do cool things like fly autonomously extra on that later.

What is drone Photogrammetric?

Photogrammetric is described as the art, science, and technological understanding of obtaining reliable documents about bodily objects and the environment through the technique of recording, measuring, and interpreting photographic and patterns of recorded radiant electromagnetic power phenomena. Advanced software program software gear have made the remaining desire more amazing due to the reality professionals can leverage these utilized sciences to create digital property with pixel statistics gathered from photographs.

Area where drones can cover

The approach has develop to be imperative for a wide variety tasks, consisting of surveying improvement internet websites and flood zones, assessing crop health and exploring fossil sites. Rescue robotic is a robotic that has been designed for the cause of rescuing people, It is used in mining accidents, town disasters, hostage situations, and explosions, It can get proper of entry to unreachable areas, It is being made with talents such as searching, reconnaissance and mapping, getting rid of or shoring up rubble, transport of supplies, medical treatment, and evacuation of casualties.

There are military Drones and how it be manipulate for a given order

Military robots are self reliant robots or remote-controlled devices designed for navy applications, limitless navy robots have been developed through skill of pretty a range armies, Marine rescue robots are underwater motors that play an fundamental feature in underwater rescuing, enterprise offshore oil and gasoline exploration, subsea works, upkeep and inspection, Marine robots can go into battle stipulations to furnish the assist to soldiers, and they can supply suppressive fire.

Can drones be use in battlefield?

Using robots on the battlefield is very useful, they trade troopers in hazardous missions such as crawling by using caves or in street-to-street metropolis combat, they restrict civilian casualties if used proper and if adequate ethical programming ought to be developed, they act as a strain multiplier; one human fighter ought to command a squad of robots working semi-autonomously.

CHAPTER TWO

COMPONENT PARTS OF DRONES

What is Drone Propellers?

You'll locate two varieties of propellers. The tractor props at the front pull the quadcopter with the aid of the air and the pusher props do exactly as discover suggests. Propellers spin in opposite guidelines to provide rise off its physics.

What is Drone Motors?

The cutting-edge drones use a brushless motor which are an approaches more surroundings pleasant than their brushed predecessors. Motor efficiently is indispensable for battery existence and surroundings pleasant flight to enable the drone to proceed to be in the air longer.

What is Drone Motor Mount?

This is now and again developed into the combination fittings with landing struts or can be section of the UAV frame, it mounts the motor.

What is Drone Landing Tools?

Most drones have a consistent landing gear. However, high-end drones will have retractable landing equipment giving a full 360 diploma view when in the air.

What is drone Booms

In many drones, the boom is part of the main body. Other drones have extended as a separate part. Shorter booms amplify maneuverability, at the same time as longer booms increase stability. Booms want to be challenging enough to maintain up in a crash, even as interfering with prop downdraft as little as possible.

What is Main Physique?

The central hub which properties the battery, foremost boards, processors, avionics, cameras, and sensors,

What is drone Electronic Speed Controllers (ESC)

Digital circuit acts to fluctuate an electric powered motor's speed, its course and perchance moreover to act dynamic brake. The ESC converts DC battery power into 3-phase AC for driving the brushless motors.

What is drone Flight Controller?

central to the whole functioning of a UAV, the flight controller regulates motor speeds, via way of ESCs, to supply steering, as well as triggering cameras or exclusive payloads;

Drone GPS module and it uses

Barring GPS region tracking, drones would have very limited uses. The GPS unit gives latitude, longitude, elevation, and compass heading.

What is drone Antenna?

For receiving inputs into the receiver unit, like any antenna,

What is drone Battery?

Battery discloses indicates electricity diploma monitoring and is integral to preserve away from crashes triggered with the resource of on foot out of juice.

Drone Gimbals

A pivoting mount, which rotates about the x, y, and z axes to furnish stabilization and manipulate of cameras or distinct sensors.

Classifications of Photographs

The Terrestrial Photograph of drone

It is the branch of photogrammetric in which is taken from a regular feature on or shut to the flooring and the picture for this motive taken is identified as terrestrial photogrammetric.

The Aerial Photograph of drone

The branch of photogrammetric in which picture is taken from a digital digicam mounted in aircraft flying over the area and the photo consequently taken is diagnosed as aerial photogrammetric.

Fields in which drones can be used

We've already stated the functions of photogrammetric in civil surveying, the consequences of which are used with the aid of ability of many entities, which consists of constructing crews, governments, establishing planners and architects.

All of the data gathered from photogrammetric informs them about the entirety from critical protection measures to achievable task results.

Drone in engineering world

In the world of engineering, drone photos helps to reflect on consideration on internet websites for construction, as good as create perspective snap photographs and 3D renderings. Engineers can produce photographs of task consequences or previews, as properly as analyze their existing day progress.

Drone estate world

In the digital age, the area 70% to 83% of millennial detect their residences on mobile devices, creating attractive, right listings can extensively decorate the purchasing for experience and their understanding of the purchase. Viewers can see the home from all angles and get a clear wondering of what they're looking out at.

Drone in military world

Photogrammetric moreover performs a role in statistics gathering for navy programs. Accurate geo-location fashions with low processing situations are integral for grasp a landscape. Aerial imagery and photogrammetric technological understanding can work jointly to create right 3D maps quickly without any human input.

Drone in medicine world

While you would maybe no longer think to put the medical region in the equal classification as land surveying, the 3D fashions that come from photogrammetric science come in available for a vary of health-related uses. It can moreover work alongside far flung sensing technological knowledge to aid improve diagnoses barring invasive procedures.

Drone in movies world

Photogrammetric can play a huge feature in set format and world-building for vary of films and video games. 3D modeling can lift unique objects to fruition in a digital world, like cityscapes for movement sequences and right historical elements, such as statues and buildings. One well-known franchise that makes use of photogrammetric is the Battlefield games, which have an art work trend that works excellent with these 3D renderings and recreations. In addition to world-building, photogrammetric can moreover assist with designing unique results and proper sets.

Drone in accounting world

Photogrammetric moreover performs a section in crime investigation. It can aid to record and measure unique statistics about a crime scene and determine what used to be bodily possible. There are moreover many photogrammetric authorities that can useful resource in the courtroom.

Drone in mining and drilling world

Project engineers and contractors can use right 3D fashions to show and structure their worksites. The information from a photogrammetric model can aid create a wise worksite with sensors and safety factors that decorate the environment. These fashions work in tandem with linked vehicles.

Drone in sports world

Analyzing athlete moves can aid coaches and researchers understand increased about their activities. They can extend digital teaching buildings and lookup about the bodily effort that gamers dissipate with the resource of monitoring their physique movements. Topographical maps moreover come in on hand for out of doorways athletes, like hikers, mountain climbers, skiers and snowboarders. Mapping a ways off areas is normally much less tough with the help of photogrammetric technology.

Drone in agriculture world

In agriculture, aerial snap shots can supply insights into soil quality, irrigation scheduling, weight-reduction plan and pests. Farmers can adjust their planting schedules or adjust irrigation and fertilizers with this information. They can moreover use photogrammetric when assessing enlarge and crop damage after storms or floods.

How drones works and it impartations

Like all things with transferring aspects and electronics, ordinary or normal use motives put on and tear. For drones, it's advocated to put into impact a things to do protection programmed on per-flight groundwork however than set dates in the calendar.

Take care of the requirements and effortless dirt and from the chassis. Anything that flies via the air will accumulate a buildup of muck, dust, bugs and pollution. Remember that you're dealing with a digital laptop so a little care is required. Don't commence interfering with electronics and circuit boards besides you are conscious of what you're doing. Even the smallest crack cans intent flight problems. Use an anti static cloth, a compressed air cleaner and a slight brush to maintain the UAV in a shiny, out-of-the-box condition.

Drones and electronic devices

Check components, take seem to be at that screws and fastenings are suitably tightened, alternatively now not over tightened as this can intent stress. You recognize the drill then again it's integral to commonly make sure the complete issue is neatly held together. Examine the motors regularly. Make sure they are clean and free from dust. Just as you strengthen acquainted with how you're vehicle runs, get to recognize how your drone sounds. Most of the noise from a drone comes from the motors.

Issues of fault in drone

If it doesn't sound right, then find out about your drone. Propellers favor to be subject to ordinary scrutiny. Faults or damage can easily lead to catastrophic penalties as quickly as the drone is in the air. Propellers prefer altering at intervals advocated through ability of the manufacturer.

Easy ways to assemble and disassemble drones

Check the propellers are free-spinning disconnect the battery first. You can additionally desire to disassemble them from the unit to pick out what the problem is if they don't rotate easily. The landing gear moreover desires a take seem to be at to make positive the UAV returns in one piece.

The closing thing you pick is to be performing a handy landing fully for something to shatter definitely as the drone makes contact with the ground. Clean out the motor chamber and study the scenario of wiring and solder joints. Make sure the antennas are free from particles to make positive an appropriate connection with the base.

Update UAV

The digital camera would possibly additionally favor a wipe over with a suitable smooth cloth and non-abrasive cleaner that you can select out up from an images retailer. Many of today's drone cameras have self-cleaning sensors, protective seals, or filters masking the sensor itself. Check that firmware and software application are up to date and going for walks the brand new release. Update mechanisms are commonly in-built to UAV constructions alternatively it will pay to make sure it hasn't been by means of threat switched off.

Complete Steps Troubleshooting

Checkmate all devices

The first step in any troubleshooting is to check for exterior damage on the drone. If there is none you can feature some vital tests to make sure your drone is definitely functional. Remove all of the propellers, use a superb battery, electrical energy up your drone and strive a flight sequence. Check if all of the motors are responding. Go by means of the drone boot sequence, the indoors assessments will take seem to be at the firmware, sensors, and camera.

Make a test

The subsequent step is to go for a convenient flight test. Don't fly too high, and try any high-speed maneuvers. If your drone is responding correctly, then you can keep flying. If this flight takes seem to be at fails, then something is damaged and there is an internal problem with your drone.

How to set drone for operations

Create Account

You favor to have software program for GitHub due to the reality that that is what Drone makes use of to get proper of entry to code, authenticate users, and add net hooks to gain events.

Have Configuration and build up

Use the opens command to authenticate runners with the most vital Drone conditions through producing a shared secret. This is the initiation of configure to assemble your Drone Server. Storing Build Logs Externally, The extent can extend greater than Gigabytes for intently used installations.

Connect server and storage facility

Though these logs are saved on the server's database, you can pick out to use exterior storage for performance, scalability, and stability. After making high quality that all the prerequisites are met, Copy the Spaces Access key to your clipboard and then change your configuration file.

How to install photogrammetric program

Point clouds outline the shapes of factors on a worksite; then again to get a photorealistic digital ground model, you desire to match these shapes with visuals.

This is the vicinity photograph stitching, or ortho-rectification, comes in. The laptop computer seems for everyday elements shared with the aid of the usage of a couple of pictures captured in the same region to go well with snap shots jointly and combine them with the element cloud. It's like a very optimal jigsaw puzzle. This gadget works suitable on most surfaces; alternatively there are obstacles to the computer's ability to apprehend patterns.

If flooring is too featureless or turbulent, like the polished domestic home windows of a developing or the churning waves of the ocean, stitching doesn't work very well. You can't go well with an attribute between photos if it's there in one photograph and lengthy previous in the next, or if each and every feature seems the equal as each other.

www.ingramcontent.com/pod-product-compliance
Lightning Source LLC
Chambersburg PA
CBHW070758220526
45467CB00014B/788